PUGMARKS AND CARBON FOOTPRINTS

A Green Humour Collection

Celebrating 35 Years of
Penguin Random House India

ADVANCE PRAISE FOR THE BOOK

'Rohan's work never fails to stop me mid-scroll. Sometimes with a smirk and at other times a belly-laugh, and always tongue-in-cheek . . . zzzzing! His Green Humour finds its mark. Rohan's work is especially important for he has been able to reach beyond just the cohort of the ecologically primed, to make important environmental news and messages accessible to the world at large'—Arati Kumar-Rao, National Geographic explorer, environmental photographer, writer and artist

'Rohan's ability to bring life to our planet's greatest challenges in a light and humorous manner connects greater audiences than the constant doom and gloom messages that are regularly shared. In my mind, he is one of our greatest assets in changing the way people engage with and understand our environmental crises. His talent and courage to speak truth to power is second to none, and this book is a testament to that'—Asha de Vos, National Geographic explorer, marine biologist and ocean educator

PUGMARKS AND CARBON FOOTPRINTS

A Green Humour Collection

ROHAN CHAKRAVARTY

VINTAGE

An imprint of Penguin Random House

VINTAGE

USA | Canada | UK | Ireland | Australia
New Zealand | India | South Africa | China

Vintage is part of the Penguin Random House group of companies
whose addresses can be found at global.penguinrandomhouse.com

Published by Penguin Random House India Pvt. Ltd
4th Floor, Capital Tower 1, MG Road,
Gurugram 122 002, Haryana, India

First published in Vintage by Penguin Random House India 2023

10 9 8 7 6 5 4 3 2 1

The views and opinions expressed in this book are the author's own and the
facts are as reported by him which have been verified to the extent possible,
and the publishers are not in any way liable for the same.

ISBN 9780143459897

Typeset in Arial by Manipal Technologies Limited, Manipal
Printed at Replika Press Pvt. Ltd, India

www.penguin.co.in

For the House Sparrows in my bird boxes
(there have been over forty clutches and over a hundred new chicks so far,
and too many to name, so here's an illustration instead)

FOREWORD

Amid the post pandemic blues, geopolitical angst and the IPCC finalizing the Synthesis Report for the 6th Assessment Report, there is a rare treat in store for the world. Rohan Chakravarty's *Pugmarks and Carbon Footprints*, a collection of over 200 of his cartoons and comics published in print and online over the last two years. For a planet recovering from the worst health emergency in a century, we need this book primarily because it puts the lens of humour on our way of looking at the natural world.

Most creative work on nature and environment is often didactic and used as social commentary, but Rohan's work is very different. His comics are deeply rooted in the behaviour and ecology of species which he brings to life with his idiosyncratic style of illustration. He then connects this to everyday events. And then he adds an elegant layer of wit or satire. This, dear reader, is a splendid combination. Firstly, it enlightens us by breaking down scientific jargon. Secondly, it turns the matter into content that is easily comprehensible across age groups. Thirdly, it engages us emotionally—whether you RoFL or LoL or Grrr. There is never a call to action in Rohan's work (unless of course commissioned to do so). So very gracefully, the colours and tones, the words and the phrases evoke a mood and a visceral response from the reader. That is the power of his work.

Until this passage was written, Rohan and I had never met. However, he was everywhere I went! He first stared at me in the form of the 'Illustrated Map of Pakke Tiger Reserve' in Arunachal Pradesh. While I knew many of the species, what I didn't know was that many of them inhabit Pakke! Next, he turned up on the wall of the NCF Interpretation Centre in Iyerpadi, Tamil Nadu, guised as the 'Valparai Landscape and Habitats' poster. I was not aware of more than half of the fifty-two animals listed on it! I read his comic plus colouring book on elephant conservation *Where's Gaju's Herd?* before I wrote a film proposal on WTI's Gaj Yatra campaign. Rohan had the audacity of entering my home as the book *Bird Business*, a gift for my children. I bragged to my son I knew the author, and I had even read his *Naturalist Ruddy—Adventurer Sleuth Mongoose* as a bedtime storybook four times over. During the pandemic, his comic strips came to my rescue when I was looking for resources for my workshops with schoolchildren. It does not end there. He pops up online across most of the resource websites I read and his work is shared by almost everyone I know over and over again!

It takes patience and practice to be a communicator for the natural world, years can pass by before you can monetize that passion and it takes immense courage to comment on contemporary events. Rohan has been consistent for thirteen years now. That is why Green Humour has consistently been among the most read and appreciated cartoon columns in India, and is making its mark around the world. That is why Rohan is a sought after storyteller and illustrator in the business of communication for the environment.

Pugmarks and Carbon Footprints is much more than a collection of Rohan's brilliant cartoons. It is a companion guide to understanding the wild, a chronicle of the current challenges to conserving nature, a commentary on contemporary politics surrounding conservation and a compendium of micro-stories. This book will compel you to laugh, yet sear through your heart, and perhaps leave you more aligned with the natural world.

Akanksha Sood Singh
National Award-winning film-maker and founder, Women of the Wild

INTRODUCTION

Dear readers,

I don't even know if I should be calling you that, because all of us are mostly 'scrollers' now. Anyway, to the ones taking the trouble of reading this in print, a hello and a warm welcome, right from where *Green Humour for a Greying Planet* left off. In case you have not read the first compilation, don't worry; this one's just a spiritual sequel, as uninspired film-makers like to call their work these days. All you need to know is that Green Humour is a series of cartoons and comics on wildlife, nature conservation, sustainability and all things green. I know fully well that cartoon compilations really don't need an unnecessary page of introduction, but I do have a few things to say.

Given the constraints that the publishing world has faced after COVID (well, look who fell in line!), we've had to bring about a few changes to this book, compared to GHFGP. The previous was a compilation of eight years of my work, and given the volume, it was distributed into eleven chapters. This time around the compilation is from two years only, and the division of chapters has been dispensed with, to keep the page count limited as well as not to disrupt your popcorn-reading routine (I don't know if GM popcorn is evolved enough yet to object to generalization). *Green Humour for a Greying Planet* was published in full colour, but *Pugmarks and Carbon Footprints* has cartoons in all their black-and-white glory like the good old times, with eight comics in colour (this isn't an attempt to sell 'retro' to today's youth, but to make the book affordable for readers nationwide). Colour-printing also has a much higher footprint than black-and-white, so please give this book some brownie points (greyscale brownies will do). It is also an era of diminishing space for cartoons in the print, and before I lose all my print columns (like I did lose a few this year), I wanted to take this opportunity to bring out a compilation with footnoted publishing information of each cartoon in the book, and feel like a professional cartoonist for once!

My rambling aside (sorry, I am well aware that every published word has a carbon footprint, including 'carbon' and 'footprint' themselves!), the cartoons in this compilation will refresh your memory of a few important environmental occurrences that have taken place in India and the world in the recent past, and introduce you to some very interesting wildlife, on land, in the air, underwater, and even in geothermal vents! If they manage to make you develop a borderline obsession for wildlife, or inspire you enough to pick up at least one sustainable personal habit, or, better still, motivate you to vote for the greenest party from whichever part of the world you are, I would consider this book to be an M.Y. Times Bestseller (my sincere apologies for the dad joke, and a few more of those that you will chance upon in this book; I am an official dad to two gorgeous Indie dogs).

All right then, go grab that popcorn (GM or otherwise), and dive right in!

Rohan Chakravarty
April 2023

The unique Spirit Bear, a resident of temperate rainforests, is revered by the First Nations people, who believe that Raven, the Creator, created the spirit bear in memory of the frozen white glaciers of the ice ages! Why exactly Spirit Bears are white in colour is still being researched, but some scientists speculate that it could be an evolutionary adaptation to catch salmon more successfully!

The Hindu Sunday Magazine, 2nd June, 2019

GEN-Z SPIDER

Sunday Mid-Day, 2nd June, 2019

MEET THE YELLOW-THROATED MARTEN

Like all weasels, the Yellow-throated Marten is among the cutest-looking animals you'll ever see. And like all weasels, Yellow-throated Martens can shock you with their killing abilities! The comic was inspired by a report of a marten taking down an adult Rhesus Macaque (an animal twice its size) in Corbett National Park.

Roundglass Sustain, 22nd July, 2019

BOOKWORM BIJAL: THE SUSTAINABLE BOOKWORM

BULK-BUYS FROM SECOND-HAND BOOKSTORES, AND NOT FROM ONLINE CONGLOMERATES.

ALWAYS CARRIES HER RUCKSACK (AND EXTRA CLOTH BAGS) ON A BOOK-SHOPPING SPREE.

BORROWS COPIOUSLY FROM FELLOW-BOOKWORMS.

OBSESSIVELY FORAGES FOR DISCARDED CARDBOARD TO MAKE BOOKMARKS.

HER KINDLE IS PERENNIALLY SET ON 'POWER-SAVER'!

PEEPS INTO THE BOOKS OF CO-PASSENGERS IN THE METRO.

STEALS FROM THE RICH, SPECIALLY THE RICH WHO HOARD & NEVER READ.

DONATES BOOKS GENEROUSLY, EVEN IF PARTING WITH THEM BREAKS HER HEART!

Pune Mirror, 28th September, 2019

In March 2020, the Red Panda was oficially split into two species (Himalayan and Chinese), thanks to geographical isolation by the Yarlang Tsangpo river and a new molecular study!

(Sikkimese translation by Dr Minla Lachungpa and Chinese translation by Ms Yuli Yang)

Roundglass Sustain, 16th March, 2020

This Earth Day, meet some weird Indian wildlife living under the earth:

The Maharashtra Caecilian
Half frog, half worm!

The Snake Skink
Half lizard, half snake!

The Horned Ghost Crab
Half crab, half alien!

The Ocellate Shieldtail Snake
Half snake, half worm!

The Swamp Eel
Half snake, half fish!

The Spiny-Tailed Lizard
Half lizard, half porcupine!

The Himalayan Mole
Half pig, half mouse!

The Indian Purple Frog
Half frog, half pig!

But full-time earth engineers... and full-time Earth Day ambassadors!

Some underground weirdos wish you a happy Earth Day!

Roundglass Sustain, 22nd April, 2020

ENDURES AN UNUSUALLY LONG GESTATION.

GIVES BIRTH UPSIDE DOWN!

FLIES CARRYING HER PUP, WEIGHING A THIRD OF HER OWN WEIGHT!

WORKS HARDER THAN THE AVERAGE BAT TO MAKE ENOUGH MILK, HUNTING A LOT MORE INSECTS!

RECOGNIZES PUP IN A COLONY OF HUNDREDS, USING ITS VOICE & SCENT.

MUMMA!

BREASTS OF STEEL: PUPS HANG ON TO THE MOTHER'S BODY BY THE NIPPLES! SOME EVEN HAVE FALSE NIPPLES FOR THIS.

THE SUPERHERO THAT NOT JUST GOTHAM, BUT THE WORLD DESERVES...

AND THE ONE IT NEEDS RIGHT NOW!

INDIA'S ENDANGERED FRESHWATER TURTLES & TORTOISES COMPLAIN ABOUT THEIR PLIGHT

THE INDIAN STAR TORTOISE

WANTED TO BE THE STAR OF SCRUB FORESTS, ENDED UP BECOMING THE STAR OF THE ILLEGAL PET TRADE.

THE ASIAN GIANT SOFTSHELL

I'VE HAD ENOUGH OF BEING A SOFT TARGET FOR POACHING & HABITAT LOSS.

RED-CROWNED ROOFED TURTLE

THANKS TO SAND MINING & DESTRUCTION OF RIVER BANKS FOR MAKING ME FEEL HOMELESS DESPITE THE ROOF OVER MY SHELL.

KEELED BOX TURTLE

ALWAYS PARANOID THAT MY SPECIES WILL END UP IN A... BOX!

THE INDIAN EYED TURTLE

CAN WE HAVE INDIA'S EYES ON TURTLE CONSERVATION NOW?

—ROHAN

Of the twenty-nine species of freshwater chelonians (turtles and tortoises) found in India, twenty-six are classified under various threatened categories on IUCN's Red List.

The Hindu Sunday Magazine, 17th May, 2020

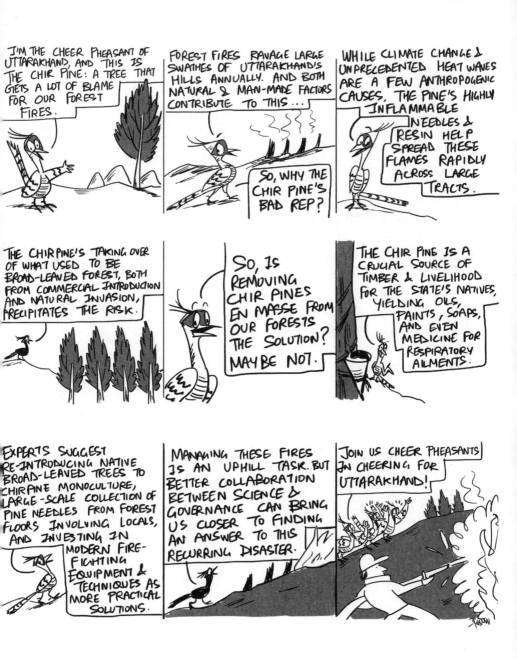

The Cheer Pheasant, a resident of Uttarakhand's Chir Pine forests, tells the complicated story about the state's annual forest fires.

Roundglass Sustain, 30th May, 2020

Diego, a legendary Galapagos Giant Tortoise, credited
with single-handedly pulling his species back from extinction thanks to his libido,
retired at the age of 100, and returned to his native island of Espanola
from the Galapagos National Park's breeding programme at Santa Cruz.

Gocomics, 20th June, 2020

THE INDIAN ENVIRONMENTAL 'MANEL'

We've all been hearing of endangered flora and fauna. But do you know of endangered ecosystems? One such habitat—the Myristica swamps—dots the Western Ghats and is now on the verge of being wiped out. Hear it from the endemic Myristica Swamp Frog, described as recently by science as 2013!

The Hindu Sunday Magazine, 5th July, 2020

Sunday Mid-Day, 5th July, 2020

Environmental clearance for the Hubbali-Ankola railway line was accompanied by falsified impact assessments and misinforation about its benefits, prompting widespread opposition.

Gocomics, 7th July, 2020

TO THE FOREST GUARD WHO DIDN'T LEAVE MY SIDE UNTIL I SWAM ASHORE...

TO THE FARMER WHO SPARED ME SOME FODDER...

TO THE TRIBAL FAMILY THAT LET A CREATURE THRICE THEIR SIZE SHARE THEIR ROOF...

TO THE VETERINARIAN WHO WADED IN WAIST-DEEP WATER TO STITCH MY WOUND...

TO ORGANIZATIONS THAT MOBILIZED FORCES AND RESOURCES EVEN BEFORE THE FLOOD HIT...

AND TO CITIZENS WHO EMPOWERED RESCUE EFFORTS WITH THEIR DONATIONS...

ASSAM RELIEF FUND
DONATE

THANKS FOR PUTTING THE "NATION" IN "NATIONAL PARK"!

—ROHAN

Roundglass Sustain, 10th July, 2020

15

~~ONCE UPON A TIME~~
EVERY SINGLE TIME I CLICK A BIRD PICTURE

Insert Ennio Morricone background track of your choice.

Gocomics, 15th July, 2020

iNaturalist is an online social network of naturalists, which helps you identify your observations and doubles up as a citizen science and biodiversity mapping project, with over a million users!

The Hindu Sunday Magazine, 19th July, 2020

CICADA KINKS

WAITING MANY YEARS FOR ACTION.

> HELLO, ADULTHOOD! LET'S DO ADULT THINGS.

STRIDULATION: A NOISY SHOW OF TYMBAL VIBRATIONS (UP TO 400 TIMES A SECOND!)

> FASTER THAN ANY OF YOUR POWER WANDS, LADIES!

ORGIESSS

DYING SHORTLY AFTER MATING.

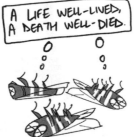

> A LIFE WELL-LIVED, A DEATH WELL-DIED.

GOLDEN SHOWERS ON BIRDWATCHERS.

> YOU LIKE THAT, HUH? HUH?

Sure, you can be kinky, but you can never be cicada-kinky!

Sunday Mid-Day, 9th August, 2020

HERE ARE SOME CONSERVATION THREATS FACED BY RAYS—

OVERFISHING FOR OUR GILL PLATES USED IN TRADITIONAL CHINESE MEDICINE...

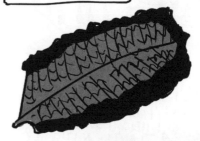

HABITAT LOSS AND DISAPPEARING CORAL REEFS...

OVER-HARVESTING FOR MAKING A TYPE OF UNTANNED LEATHER CALLED SHAGREEN.

MUCH TO MY CHAGRIN.

Roundglass Sustain, 9th August, 2020

Jammu and Kashmir's state animal, the Hangul asks some 'anti-national' questions.

The Hindu Sunday Magazine, 16th August, 2020

NAG NAALA, AFTER WHICH NAGPUR IS NAMED, ONCE USED TO BE THE RESPLENDENT NAG RIVER.

ITS POINT OF ORIGIN, THE LAVA HILLS, HAVE BEEN BLASTED FOR DEVELOPMENT, AND THE VERY FACT THAT THE RIVER ORIGINATED HERE HAS BEEN ERASED FROM NAGPUR'S RECORDS AND MEMORY.

TODAY, THE RIVER IS OFICIALLY SAID TO ORIGINATE FROM THE AMBAZARI DAM, A REAL ESTATE TOP-SPOT.

MY MY NAGPUR

SUCCESSIVE GOVERNMENTS HAVE FAILED TO REVIVE THE RIVER, WHICH GRAPPLES WITH SEWAGE, URBAN WASTE, FLOODING, CONGESTION, POLLUTION & A GROWING HUMAN POPULATION...

WHILE THE RIVER'S ORIGINS, NATURAL HISTORY, ECOLOGY & HERITAGE LIE CRUSHED & FORGOTTEN UNDER THE EPONYMOUS CITY.

THE NAG RIVER GAVE NAGPUR ITS IDENTITY, AND NAGPUR GAVE THE RIVER AN IDENTITY-CRISIS.

A Nag (Spectacled Cobra), the municipal symbol of my hometown Nagpur, tells you the story of the river this city was named after.

Sunday Mid-Day, 6th September, 2020

21

POTTER WASPS RECREATE THE POTTERY SCENE FROM 'GHOST' AND BUILD SOMETHING OF ACTUAL UTILITY.

'Ghost' starring Potter Wasps.

Roundglass Susitain, 22nd September, 2020

World Migratory Bird Day campaign comic for Birdlife International and the
East Asian Australasian Flyway Partnership, 8th October, 2020 (figures adjusted
after Bar-tailed Godwits recently broke their own record of 11000 km when a satellite-
tagged individual was recorded flying 13500 km non-stop, in October, 2022!)

The comic was created with inputs from Ms Frauke Quader of The Society to Save Rocks, which has been campaigning to save Hyderabad's remaining rock formations from unplanned development and frequently conducts rock walks in the city.

The Hindu Sunday Magazine, 19th October, 2020

The Hindu Sunday Magazine, 2nd January, 2022

WHAT CLIMATE CHANGE MEANS TO DIFFERENT PEOPLE

Niti Aayog's proposal to build a megacity in the Little Andaman Island
violated provisions of the Indian Forest Act, Andaman and Nicobar Protection
of Aboriginal Tribes Regulation and CRZ rules, endangering the Onge Tribal Reserve.

Gocomics, 4th February, 2021

Valentine's Day selfie tutorials from dragonflies.

Sunday Mid-Day, 14th February, 2021

The rare Lilac Silverline butterfly, a resident of Hesarghatta grasslands in Bangalore, comments on the government's decision to ignore citizens' plea for giving Hesarghatta a protected status and instead earmark it for developing a film city.

The Hindu Sunday Magazine, 14th February, 2021

Sunday Mid-Day, 16th February, 2021

The Hindu Sunday Magazine, 20th February, 2021

With great flight comes great responsibiity: one which fruit bats perform like no other mammal! Being the only mammals capable of true flight, their wings ensure that they spread their seeds of labour far and wide, often being responsible for the regenartion of entire rainforests.

Roundglass Sustain, 26th February, 2021

Sunday Mid-Day, 28th February, 2021

An Indian Pangolin tries his hand at political cartooning.

The Hindu Sunday Magazine, 28th February, 2021

There are many reasons why raptors are my favourite wildlife. Here's one.

A Women's Day Special,
The Hindu Sunday Magazine, 8th March, 2021

ORCA WINFREY INTERVIEWS MEGHAN MACKEREL.

FIELD CHARACTERISTICS OF THE HUMP-NOSED PIT VIPER

GREYISH-PINK COLOUR WITH DARK UNDERSIDE THAT ENHANCES CAMOUFLAGE IN LEAF LITTER

SKIN MOTTLED WITH DARK TRIANGULAR BLOTCHES

TRIANGULAR HEAD WITH A PALE LINE ALONG THE SIDES

UPTURNED SNOUT FROM CONSTANTLY TURNING ITS NOSE UP AT INDIAN NEWS CHANNELS.

The Hindu Sunday Magazine, 14th March, 2021

SOME SIMILARITIES BETWEEN THE SAND BUBBLER CRAB AND BILL WATTERSON

LEGENDARY ARTISTS, WAY AHEAD OF THEIR TIME

FAITHFUL TO TRADITIONAL MEDIA EVEN IN THE FACE OF DIGITAL ONSLAUGHT

HAVE BETTER ENDEAVOURS TO INVEST THEIR TIME IN THAN SOCIAL MEDIA

COMPLETE CONTEMPT FOR THE PAPARAZZI

SLAM!

HAVE IMAGINARY FRIENDS

GIVE BUBBLES A WHOLE NEW PURPOSE AND MEANING

HAVE PARTICULARLY SHAMELESS PREDATORS

(BOOTLEGGED C&H MERCHANDISE)

RETIRE AT THE PEAK OF POPULARITY

A World Cartoonists' Day special.

Roundglass Sustain, 5th May, 2021

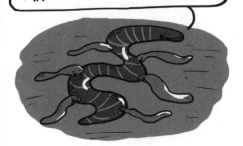

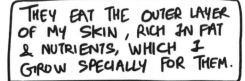

Matriphagy meets Mother's Day.

The Hindu Sunday Magazine, 9th May, 2021

SOME PORTFOLIOS HELD BY THE MUD-DAUBER WASP MOM

ARCHITECT: CONSTRUCTS NESTS WITH NEAT, CYLINDRICAL CELLS FOR HER LARVAE.

SPECIALIZED POTTERY: USES CLAY BALLS TO BUILD HER MASTERPIECE.

ENGINEER: HER NESTS HAVE UP TO 25 CELLS EXCLUSIVE TO INDIVIDUAL LARVAE.

HOME DECOR: USES BIRD DROPPINGS AND MOSS TO CONCEAL THE NEST FROM PREDATORS.

SERIAL KILLER: PROVISIONS EACH OF HER LARVAE WITH A PARALYZED SPIDER TO EAT.

SUPERMODEL: I MEAN, JUST LOOK AT THOSE COLOURS!

THOUGHT I'D SUMMARIZE THAT ON MY BUSINESS CARD.

BLACK & YELLOW MUD DAUBER

MOTHERHOOD | MULTITASKING | MURDER.

—RMNN

A Mother's Day special.

Sunday Mid-Day, 9th May, 2021

One of the many threatened species whose nesting grounds will be wiped out for the Great Nicobar Transshipment Project is the Leatherback Turtle.

Sunday Mid-Day, 16th May, 2021

NATURE TV BRINGS TO YOU THE NEWEST MAMMAL DISCOVERED IN INDIA - THE NARCONDAM SHREW, NATIVE TO THE NARCONDAM ISLAND IN THE ANDAMANS!

AFTER HIDING FROM SCIENCE ALL THIS WHILE, WHAT MADE YOU SHOW UP FINALLY?

THE VOLCANOES OF THIS IMPERILED ISLAND MAY BE DORMANT, BUT I WANT INDIA'S INTEREST IN ITS ECOLOGY & CONSERVATION TO ERUPT NOW!

SHREWWWD!

Meet the newest mammal discovered from India, an insectivore endemic to the remote Narcondam Island: the Narcondam Shrew!

Sunday Mid-Day, 16th May, 2021

Inspired by a few instances of wildlife actually using resorts in lockdown for a quick retreat, during the COVID lockdown!

The Hindu Sunday Magazine, 23rd May, 2021

The Red-eared Slider, an aquarium trade favourite, is an invasive species that is wreaking havoc on Indian turtles and their habitats, thanks to the pet trade. You can help by discouraging the trade in the species (and any other turtle/tortoise), and by NOT releasing a Red-eared Slider into a natural space. You can also help by reporting wild sightings of this species on the India Biodiversity Portal, which researchers from the group 'Freshwater Turtles and Tortoises of India' are keeping a track of.

Sunday Mid-Day, 23rd May, 2021

On Amazonian palm oil's double standards, which, despite being introduced with a promise of sustainability, has been taking a toll on wild habitat and tribal land with deforestation and chemical pollution.

The Hindu Sunday Magazine, 30th May, 2021

India's scorpions, especially our endemic, rock-dwelling varieties, are under increasing threat from habitat loss and the pet trade.

Sunday Mid-Day, 6th June, 2021

Roundglass Sustain, 13th June, 2021

A short story of sex and the Antechinus, an Australian mouse-like marsupial.

Roundglass Sustain, 16th June, 2021

The forests of Buxwaha in Madhya Pradesh, an important wildlife corridor, could be lost to a diamond mine being planned by the Aditya Birla group in the region, threatening not just endangered wildlife but also tribal livelihoods.

Chakmak, 18th June, 2021

The Hindu Sunday Magazine, 20th June, 2021

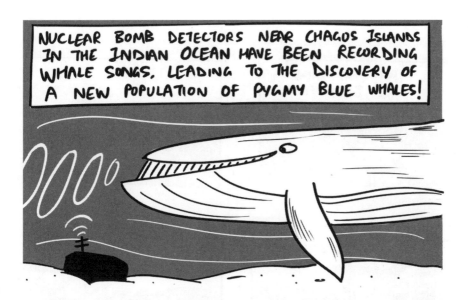

Nuclear bomb detectors placed near Chagos Islands detected whale sounds, which when analysed by a group of scientists from the University of New South Whales, led to the discovery of a new population of Pygmy Blue Whales in the Indian Ocean!

Sunday Magazine, 20th June, 2021

The Sri Lankan coastline was ravaged by nurdle pollution after the vessel MV Xpress Pearl had a fire accident, resulting in the death of thousands of marine animals. Sri Lankan marine biologist Asha De Vos and her organization Oceanswell.org launched a web portal 'Nurdle Tracker', where volunteers could upload data on nurdles to help cleaning efforts.

Sunday Mid-Day, 4th July, 2021

Inspired by an article titled 'Shifting our Gaze' written for *The Wire Science*, by Bidyut Sarania, Krishnapriya Tamma, Samira Agnihotri, Subhashini Krishnan and Sutirtha Lahiri.

The Hindu's Sunday Magazine, 4th July, 2021

The Hindu Sunday Magazine, 11th July, 2021

BATS & BOOZE GO BACK A LONG WAY!

YOU MIGHT KNOW FROM VIRAL WEB ARTICLES THAT THE MEXICAN LONG-TONGUED BAT POLLINATES AGAVE FLOWERS THAT YIELD TEQUILA...

WHAT YOU MIGHT NOT KNOW IS THAT INDIA'S VERY OWN SHORT-NOSED FRUIT BATS POLLINATE & DISPERSE THE SEEDS OF LIQUOR-YIELDING TREES LIKE MAHUA!

IN FACT, MAHUA'S NIGHT-BLOOMING FLOWERS & FRUITS HAVE EVOLVED TO BE POLLINATED & DISPERSED SPECIALLY BY FRUIT BATS!

MAHUA LIQUOR HAS BEEN CENTRAL TO INDIGENOUS CULTURES, AND IS NOW BEGINNING TO GO MAINSTREAM, WITH EVEN THE GOVERNMENT TRYING TO CASH IN ON IT!

GOSH. THIS COUNTRY OWES ME SO MUCH MORE THAN A BAD REP!

INCLUDING THE NEXT TOAST YOU RAISE.

TO BATS.

Roundglass Sustain, 18th July, 2021

AMAZING DEFENCES OF THE INDIAN MOON MOTH

CATERPILLARS EMIT LOUD CLICKS
AND REGURGITATE A VILE FLUID

ADULTS HAVE LEAF-LIKE
COLOURATION FOR CAMOUFLAGE

EYESPOTS ON THE WINGS
DETER PREDATORS

THE EXTENDED TAIL FRILLS
DEFLECT BAT SONAR!

A National Moth Week special.

Roundglass Sustain, 24th July, 2021

56

THIS WEEK'S ENVIRONMENTAL HEADLINES:

FLOODS DEVASTATE THE HENAN PROVINCE OF CHINA...

EUROPE'S FLOOD DEATH TOLL CONTINUES TO RISE...

WILDFIRES RAGE ACROSS SIBERIA...

SEVERE HEAT WAVES CONTINUE IN PARTS OF THE USA, CANADA & SCANDINAVIA...

UNUSUALLY HEAVY RAINS & FLOODS DISRUPT LIFE IN VARIOUS PARTS OF INDIA...

THE BILLIONAIRE SPACE RACE ADDS A STAGGERING LOAD OF EMISSIONS TO THE ATMOSPHERE...

FOSSIL FUEL GIANTS CONTINUE RECEIVING SUBSIDIES FROM THE WORLD'S BIGGEST GOVERNMENTS...

ZEROWASTE ZINDA HAS SET OUT TO SAVE THE PLANET WITH HER BAMBOO TOOTHBRUSH.

ROHAN

Sunday Mid-Day, 25th July, 2021

A World Tiger Day special.

Roundglass Sustain, 29th July, 2021

Sunday Mid-Day, 1st August, 2021

Animal welfare often comes in direct conflict with wildlife conservation, and one such grey space is the issue of feral dogs in India. Their growing numbers are connected with municipal management in towns, cities and their outskirts, and they pose a direct threat to wildlife conservation. Trust India's leopards to come to the rescue, until the other lobbies figure a solution out!

Roundglass Sustain, 6th August, 2021

ELEPHANT CONSERVATION THE FIRST WORLD WAY

200 YEARS LATER:

On UK's Aspinall Foundation's plans of sending captive elephants back to Africa.

The Hindu Sunday Magazine, 8th August, 2021

Roundglass Sustain, 14th August, 2021

SOME BENEFITS OF SEAWEED CULTURE

AMONG THE LOWEST CARBON GENERATING FOOD ITEMS IN THE WORLD.

REQUIRES NO ARABLE LAND. CAN BE FARMED AS NATURAL SEAWEED FORESTS.

AN ECOSYSTEM FOR MARINE ORGANISMS

AN IODINE-RICH SOURCE OF NUTRITION

HELPS STORE CARBON AND COMBAT OCEAN ACIDIFICATION & DEOXYGENATION

HAS THE POTENTIAL TO TRANSFORM THE SEAFOOD INDUSTRY FROM AN EXTRACTIVE TO A RESTORATIVE ONE.

—RAWW

SAVE THE TIGER, BREED HOUSE CATS.

IF THAT SEEMED RIDICULOUS TO YOU, I HAVE SOME NEWS...

THAT'S EXACTLY WHAT MOST BEE CONSERVATION INITIATIVES & BEEKEEPERS END UP DOING BY REARING EUROPEAN HONEY BEES INSTEAD OF CONSERVING INDIGENOUS SPECIES!

AND THAT'S A MAJOR PROBLEM BEES IN INDIA FACE TODAY, OTHER THAN HABITAT LOSS, CLIMATE CHANGE & INSECTICIDES.

INDIGENOUS BEES POLLINATE A VARIETY OF INDIGENOUS FLORA, UNLIKE THE EXPENSIVE AND UNPREDICTABLE EUROPEAN IMPORTS. PRESERVING WILD BEES MEANS PRESERVING AN ECOSYSTEM!

WHILE APICULTURE AND BEE CONSERVATION NEED TO FOCUS ON INDIGENOUS BEES, YOU CAN CONTRIBUTE BY SUPPORTING SMALLER COLLECTIVES THAT PRODUCE INDIGENOUS HONEY, OVER INDUSTRIAL HONEY BRANDS.

TRUST ME, THE ADDED SWEETNESS OF SAVING A SPECIES TAKES THE TASTE OF HONEY TO ANOTHER LEVEL!

SMMMACK!

Sunday Mid-Day, 22nd August, 2021

64

THE MALE WATER STRIDER COURTS A FEMALE

Bollywood can often get quite real in nature, especially when you're a water strider.

Roundglass Sustain, 28th August, 2021

I HAVE A QUESTION.

YES?

I KNOW YOU HAVE POOR EYESIGHT & HUNT USING ECHOLOCATION. BUT HOW DID YOU MAKE IT LOOK SO QUICK & EASY?

I MEAN, DOESN'T SONAR INVOLVE MIND-BOGGLING MATHEMATICAL EQUATIONS & SOPHISTICATED TRIGONOMETRY? HOW DID YOU DO THAT IN A SNAP?!

WELL, MATH WAS MY FAVOURITE SUBJECT IN SCHOOL!

NO WONDER THESE WEIRDOS HANG UPSIDE DOWN ALL THE TIME.

Have you ever seen a bat hunt an insect down in flight and wondered how the hell did that look so easy?

Sunday Mid-Day, 28th June, 2021

AREN'T YOU A MOON WRASSE, CAPABLE OF CHANGING YOUR SEX?

YES.

SO ARE YOU A GIRL OR A GUY RIGHT NOW?

WELL, I DON'T REALLY IDENTIFY WITH ANY SPECIFIC GENDER KNOWN TO THE ENGLISH LANGUAGE.

HUH?

LOOK, MOON WRASSES ARE SEQUENTIAL HERMAPHRODITES. I BRED AS A FEMALE FIRST, AND AM NOW IN THE PROCESS OF CHANGING INTO A MALE.

AH, SO YOU'RE ONE OF THOSE PEOPLE WHO KEEP USING FUNNY PRONOUNS!

FUNNY PRONOUNS? A SIMPLE "THEY" OR "THEM" WOULD SUFFICE.

THEY?! HAHAHA! THAT'S NOT EVEN GRAMMATICALLY CORRECT! HOW DO YOU EVEN USE THAT IN A SENTENCE?

LET ME GIVE YOU AN EXAMPLE.

"A MOON WRASSE ATE A PESKY POLYCHETE AFTER **THEY** GOT IRRITATED WITH HIS INEPTITUDE."

Sunday Mid-Day, 29th August, 2021

As researchers flee the Taliban rule, Afghanistan's national animal,
the Snow Leopard ponders the future of science and conservation in this beautiful country.

The Hindu Sunday Magazine, 29th August, 2021

Australia's Tammar Wallaby is one of the many animals capable of arresting its pregnancy, a feat shared by other marsupials, bears, rodents, armadillos and weasels.

Sunday Mid-Day, 5th September, 2021

THE MALABAR WHISTLING THRUSH COMPLAINS

Roundglass Sustain, 9th September, 2021

Roundglass Sustain, 10th September, 2021

NATURE BOOK REVIEWS BY WILD ANIMALS

WILD AND WILFUL
Neha Sinha

CITIES AND CANOPIES
Harini Nagendra
Seema Mundoli

EVERY CREATURE HAS A STORY
JANAKI LENIN

"Personal, beautiful & concise. Glad I could finish reading before going extinct!"
—The Great Indian Bustard

"A lesson or two for every reader; even a learned seed-disperser like me."
—The Indian Flying Fox

"Now every creature has a fine P.R. agent: The author of "MY HUSBAND..." or should I say "OUR HUSBAND & OTHER ANIMALS"?
—The King Cobra

HOW I BECAME A TREE
Sumana Roy

THE VANISHING
India's Wildlife Crisis
PRERNA SINGH BINDRA

"Nobody does it better than Sumana. Not even me!"
—The Bark Gecko

"Who put this in 'Non-fiction'?! This should be in the 'horror' section!"
—The Indian Elephant

Sunday Mid-Day, 12th September, 2021

The Hindu Sunday Magazine, 13th September, 2021

The Hindu Sunday Magazine, 19th September, 2021

I'M THE ENDANGERED SEA PANGOLIN, A DEEP SEA SNAIL COVERED IN AN IRON-SULFIDE SHELL, LIVING AROUND HYDROTHERMAL VENTS IN THE INDIAN OCEAN.

AND THESE ARE POLYMETALLIC SEA NODULES RICH IN METALS USED TO MAKE ELECTRONICS & ELECTRIC CAR BATTERIES, WHICH HAVE NOW CAUGHT THE FANCY OF DEEP SEA MINING COMPANIES.

WHILE DEEP SEA MINING IS BEING TOUTED AS A SOLUTION TO CLIMATE CHANGE, MARINE BIOLOGISTS PREDICT THAT THIS COULD WIPE OUT UNKNOWN SPECIES AND ECOSYSTEMS, CAUSING UNPRECEDENTED CASCADE EFFECTS ON THE PLANET.

WILL HUMANITY SAVE THIS DEEP SEA IRON MAN, OR LET MINERS PLUNDER OUR "INFINITY STONES"?

IS IT BETTER TO BE UNDISTURBED OR ALIVE, OR IS IT TOO MUCH TO ASK FOR BOTH?

Sunday Mid-Day, 19th September, 2021

LIFE WITH A STRANGLER FIG

Roundglass Sustain, 21st September, 2021

VISIT FAROE ISLANDS

A TRAVEL ITINERARY FOR DOLPHINS

ARE YOU A PILOT WHALE? COME CRASH AT THE FAROES...

(LITERALLY)

ARE YOU AN ATLANTIC WHITE-SIDED DOLPHIN?

COME, WITNESS OUR DARK SIDE!

FEEL THE THRILL OF BEING CHASED BY FISHING VESSELS AND DRIVEN TO THE SHORE!

EXPERIENCE THE BLOW OF THE SPECIAL SPINAL LANCE BY OUR TRAINED DOLPHIN HACKERS!

GET SLAUGHTERED WITH YOUR ENTIRE POD: A COMPLETE FAMILY EXPERIENCE!

YOU'RE NOT JUST MEAT FOR THIS RICH EUROPEAN COMMUNITY. YOU'RE FAROESE TRADITION!

SO GET AWAY FROM THE GRIND AND JOIN US FOR THE GRINDADRÁP!

ROMAN

On the mindless slaughter of Long-finned Pilot Whales and Atlantic White-sided Dolphins in the Faroe Islands in the name of 'tradition'.

Sunday Mid-Day, 26th September, 2021

The recent declaration of the extinction of the Ivory-billed Woodpecker made an Indian woodpecker species, the Great Slaty Woodpecker, the largest wild woodpecker in the world.

The Hindu Sunday Magazine, 10th October, 2021

FOREST CONSERVATION ACT AMENDMENT

Meet Cylix tupareomanaia, a Pygmy Pipehorse from New Zealand:
the world's first creature to be officially named in collaboration
with an indigenous tribe, the Ngatiwai. The species name in Maori translates to 'garland of the seahorse'.

The Hindu Sunday Magazine, 17th October, 2021

The Hindu Sunday Magazine, 24th October, 2021

Cartoonstock Cartoonathon at the Cop26 (Glasgow), 31st October, 2021

Cartoonstock Cartoonathon at the Cop26 (Glasgow), 31st October, 2021

As details of the menu for the COP26 Glasgow delegates made the news, a severe drought and famine raged across Madagascar.

Sunday Mid-Day, 8th November, 2021

Odisha's famed 'black tigers' from Simlipal Tiger Reserve have long harboured a genetic secret, finally decoded by the eminent biologist Dr Uma Ramakrishnan and Vinay Sagar from the NCBS. The genetic mutation is linked to inbreeding and restricted gene flow.

Roundglass Sustain, 9th November, 2021

SOME EVOLUTIONARY MARVELS RESULTING FROM GEOGRAPHICAL GAPS IN THE WESTERN GHATS

THE NILGIRI SHOLAKILI

THE WHITE-BELLIED SHOLAKILI

ONCE THOUGHT TO BE THE SAME SPECIES; NOW CLASSIFIED AS DISTINCT SPECIES NORTH & SOUTH OF THE PALGHAT GAP.

THE PALANI CHILAPPAN

THE ASHAMBU CHILAPPAN
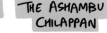

ONCE CONSIDERED THE SAME SPECIES; NOW RECOG- -NIZED AS DISTINCT SPECIES NORTH & SOUTH OF THE ACHANKOVIL GAP.

THE NILGIRI CHILAPPAN

THE BANASURA CHILAPPAN

ONCE THOUGHT TO BE THE SAME SPECIES; NOW RECOGNIZED AS DIFFERENT SPECIES OCCUPYING SKY ISLANDS AT DIFFERENT ELEVATIONS.

THE PERROTET'S WOOD SNAKE

THE ANAMALAI WOOD SNAKE

ONCE CONSIDERED THE SAME SNAKE; NOW SPLIT INTO TWO DISTINCT SPECIES NORTH & SOUTH OF THE PALGHAT GAP.

WHO THOUGHT GAPS WITHIN MOUNTAINS COULD HELP BRIDGE GAPS IN OUR UNDERSTANDING OF EVOLUTION!

-ROHAN

The Hindu Sunday Magazine, 3rd May, 2020

LGBTQ+ WILDLIFE

L : NEW MEXICAN WHIPTAIL LIZARD :

AN ALL-FEMALE SPECIES. FEMALES REPRODUCE BY PARTHENOGENESIS, OFTEN SIMULATING SEX TO IMPROVE FERTILITY.

G : GIRAFFES :

MALES OFTEN CARESS EACH OTHER AFTER NECKING, LEADING UP TO MOUNTING.

B : BONOBOS :

BOTH SEXES MATE WITH EITHER SEX, OFTEN TO RESOLVE CONFLICTS.

T : REEF FISH :

MOON WRASSE: STARTS LIFE AS FEMALE. CHANGES SEX TO MALE LATER.

PACIFIC ANEMONEFISH: BREEDING MALES CAN CHANGE SEX TO FEMALE.

Q : OCTOPOTEUTHIS SQUID :

HAS DONE AWAY WITH ALL GENDER PREFERENCES TO FIND THE RARE MATE IN THE DEEP SEA.

+ : BRAHMINY BLINDSNAKE :

A SOLITARY, ALL-FEMALE SPECIES. REPRODUCES BY PARTHENOGENESIS, AND IS VERY CONTENT WITHOUT A MATE.

COME ON, NOW. LOWER THAT RAISED EYEBROW ALREADY. THE RAINBOW SHINES ACROSS THE ENTIRE ANIMAL KINGDOM!

— Rawn

A Pride Month special

Sunday Mid-Day, 14th June, 2020

THE SAHYADRI BLUE OAKLEAF BUTTERFLY'S WORK-FROM-HOME GET-UP

THE SAHYADRI BLUE OAKLEAF BUTTERFLY'S LAVANI COSTUME

Meet a very special butterfly from my home state, Maharashtra, which dazzles from above, and resembles a dried leaf from below!

Sunday Mid-Day, 17th September, 2020

Sunday Mid-Day, 18th January, 2021

A GLACIER MELTS IN THE UPPER HIMALAYAS...

CAUSING WARMER WINDS TO BLOW INTO THE ARABIAN SEA...

BIOLUMINESCENT DINOFLAGELLATES CALLED 'SEA SPARKLES' BLOOM, FACILITATED BY THE WARMTH...

AS THE SPARKLES PROLIFERATE, THEY DESTROY THE DIATOMS THAT MARINE LIFE DEPENDS ON, AND GATHER AMMONIA, CREATING A TOXIC ENVIRONMENT.

AND THAT IS HOW THE MELTING OF HIMALAYAN ICE IS CAUSING A COLLAPSE OF FISHERIES IN THE ARABIAN SEA.

WHAT A SPARKLING TIME TO BE ALIVE!

Sunday Mid-Day, 1st February, 2021

HIMALAYAN WILDLIFE CELEBRATES THE RHODODENDRON BLOOM, BY SPORTING RED:

THE BLOOD PHEASANT

THE RED PANDA

THE SCARLET FINCH

THE SHORT-BILLED MINIVET

THE BHUTAN GLORY BUTTERFLY

THE CRIMSON SUNBIRD

GOSH, IS THAT SMILE CONTAGIOUS, LADY!

Roundglass Sustain, 29th March, 2021

Skin Positivity with Crocs

DARK

BLACK CAIMAN

FAIR

ORINOCO CROCODILE

SPECKLED

CUBAN CROCODILE

BUMP ON THE NOSE

GHARIAL

PATCHY

NILE CROCODILE

DOTTED WITH OSTEODERMS

SPECTACLED CAIMAN

BLEACHED ORANGE FROM BAT GUANO IN CAVE WATERS

AFRICAN DWARF CROCODILE

-RMN

Sunday Mid-Day, 22nd August, 2021

Sikkim's State Animal, the Red Panda, announces Sikkim's newly chosen official State Butterfly, the Blue Duke!

The Hindu Sunday Magazine, 10th July, 2022

NATURE TV BRINGS TO YOU INDIA'S ONLY KNOWN PITCHER PLANT— THE KHASI PITCHER FROM MEGHALAYA! INTERESTINGLY, THIS PITCHER USES U.V. LUMINESCENCE TO ATTRACT PREY!

LET'S FIND OUT HOW THIS PLANT ACHIEVES ITS EXTRAORDINARY GLOW...

WOAH!

SLIP

NOOOO!

GULP.

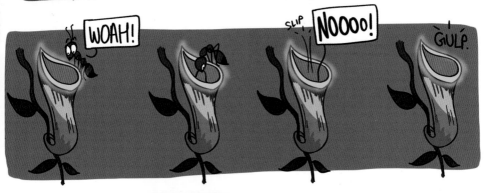

YOU REALLY THOUGHT I'D LET YOU REVEAL MY SKINCARE SECRETS TO THE WORLD SO EASILY, DARLING?

—ROHAN

Roundglass Sustain, 3rd May, 2022

The Hindu Sunday Magazine, 22nd May, 2022

KNOWN FOR THEIR EXQUISITE COLOURS AND PATTERNS, NUDIBRANCHS ARE SOFT-BODIED MARINE MOLLUSCS. THEY SHED THEIR SHELLS AFTER THE LARVAL STAGE.

BYE BYE, SHELL! TIME TO EMBRACE MY INNER NUDIE!

THE WORD 'NUDIBRANCH' MEANS 'NAKED GILLS'!

nudie_&_proudie

#free the gills

NUDIBRANCHS ARE FOUND IN SEAS WORLDWIDE, FROM INTERTIDAL TO PELAGIC DEPTHS. THEIR GREATEST DIVERSITY OCCURS IN WARM, SHALLOW REEFS.

BECAUSE EVERYTHING SASSY BELONGS TO THE TROPICS!

NUDIBRANCHS HAVE GENITALS ON THE RIGHT SIDE OF THEIR BODIES!

WHAT'S THIS SEX POSITION CALLED?

THE AWKWARD HANDSHAKE.

ALL KNOWN NUDIBRANCHS ARE CARNIVORES. THEY EAT MARINE INVERTEBRATES LIKE SPONGES, TUNICATES, JELLYFISH & ANEMONE.

WELL, THE MEAT I EAT LOOKS LIKE PLANTS, SO I STILL QUALIFY FOR VEGANUARY.

NUDIBRANCHS HAVE RIBBON-LIKE TONGUES COVERED IN TINY TEETH CALLED RADULA, WHICH HELP THEM DIGEST THEIR PREY.

RAD, RIGHT?

BEING SOFT-BODIED, NUDIBRANCHS RELY ON EXTERNAL DEFENCE MECHANISMS. SOME DERIVE TOXINS FROM THE STINGING CELLS OF THE ANEMONE THEY EAT!

A FASHIONISTA AND AN ARMS SMUGGLER. BOY, HOLLYWOOD IS GOING TO WORSHIP ME!

NUDIBRANCHS BREATHE THROUGH 'CERATA'— HORN-LIKE PROJECTIONS ON THEIR BODIES. THESE ALSO HOUSE STOLEN STINGING CELLS FROM HUNTED ANEMONE, THAT NUDIBRANCHS USE FOR THEIR OWN DEFENCE.

UNCOMFORTABLE TRUTH: TOXICITY SOMEHOW ALWAYS ENHANCES SEX APPEAL.

BEING CARNIVORES, NUDIBRANCHS ARE IMPORTANT INDICATORS OF THE HEALTH OF A REEF ECOSYSTEM.

ISN'T IT A BONUS WHEN YOUR BRAND AMBASSADOR IS ALSO A STYLE ICON?

7 REASONS TO DIM YOUR LIGHTS FOR BIRDS

1. MOST MIGRATORY BIRDS USE THE COVER OF DARKNESS FOR A SAFE FLIGHT. LIGHT POLLUTION CAN CAUSE US TO STRAY.

2. LIGHT POLLUTION INCREASES THE RISK OF COLLISIONS.

3. LIGHT POLLUTION DISRUPTS OUR BIOLOGICAL RHYTHM.

4. LIGHT POLLUTION IS A THREAT TO MANY OTHER WILD ANIMALS, LIKE INSECTS, NESTING SEA TURTLES AND THEIR HATCHLINGS.

5. LIGHT POLLUTION IS LINKED TO HUMAN SLEEP DISORDERS AND HEALTH PROBLEMS.

6. OUR RADIANCE IS ENOUGH TO BRIGHTEN UP YOUR LIVES!

7. YOU'RE ALL BRIGHT ENOUGH TO UNDERSTAND THAT WE NEED A DIM NIGHT SKY, AREN'T YOU?

World Migratory Bird Day campaign comic for Birdlife International and the East Asian–Australasian Flyway Project, 8th October, 2022

The Hindu Sunday Magazine, 11th December, 2022

MAY YOUR NEW YEAR BE LIKE THE GOLDEN ORIOLE!

A FLASH OF JOLLY YELLOW AMIDST GLOOMY GREY

JOY IN EVERY CUP

ENOUGH CHEER WITHIN YOU TO FILL THE NEIGHBOURHOOD WITH

MAKING THE MOST EVEN OUT OF WEEDS

SCRUBBING OFF EVERY THORN IN YOUR WAY

A REDEFINING YEAR FOR YOUR STYLE STATEMENT

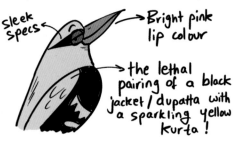

The Hindu Sunday Magazine, 1st January, 2023

PHOSPHORUS-DRIVEN ALGAL BLOOMS ARE ACCELERATING THE MELTING OF ICE SHEETS IN GREENLAND.

GROWING UP IN THE MAPUCHE COMMUNITY OF INDIGENOUS SUBSISTENCE FARMERS, YOUNG ELISA LONCÓN WOULD FEAST ON WILD BERRIES & FRUITS IN THE CHILEAN COUNTRYSIDE...

TODAY, THOSE WILD BERRIES & FRUITS ARE OVERRIDDEN BY INDUSTRIAL EUCALYPTUS AND PINE FARMS THAT PARCH THE LANDSCAPE...

BUT TODAY, ELISA IS ALSO LEADING A CONVENTION TO REWRITE THE CHILEAN CONSTITUTION, REVOLVING IT AROUND INDIGENOUS PRINCIPLES AND ECOSYSTEM REVIVAL.

THAT WAS SOME SPLENDID SEED DISPERSAL, TEAM!

On Elisa Loncón, an indigenous Mapuche political leader,
leading the convention to rewrite Chile's constitution with indigenous principles at its centre.

The Hindu Sunday Magazine, 14th November, 2021

Roundglass Sustain, 22nd November, 2021

NO, THIS ISN'T A GOATIE. THIS IS THE LARVA OF AN EPOMIS BEETLE!

IN A TOTAL TURNING OF THE PREDATOR-PREY TABLES, THIS LITTLE INSECT PREYS ON ITS OWN PREDATORS: FROGS!

THE ADULT BEETLES SYNCHRONIZE THEIR BREEDING CYCLES WITH OURS, SO THAT THEIR LARVAE WILL HAVE YOUNG FROGS TO PREY ON.

FIRST THE LARVAE LURE US WITH WIGGLY MOVEMENTS OF THEIR ANTENNAE, PRETENDING TO BE EASY PREY...

AND BEFORE WE KNOW IT, THEY'VE CAUGHT HOLD OF OUR THROATS! NOW THIS PARASITE WILL CLING ON TO ME UNTIL IT HAS SUCKED ALL THE LIVING LIFE OUT OF MY BODY!

I'D TRADE THIS THING FOR YOUR CLINGY EX ANY DARN DAY.

-RAMAN

Roundglass Sustain, 22nd November, 2021

DELHI's DAILY DILEMMAS

Sunday Mid-Day, 2nd December, 2021

7 SIGNS THAT YOU'RE A BUG-WATCHER

WHICH BUTTERFLY FAMILY IS YOUR BALDING PATTERN?

HEDYLIDAE
(AMERICAN MOTH-BUTTERFLIES)

PAPILLIONIDAE
(SWALLOWTAILS)

HESPERIIDAE
(SKIPPERS)

PIERIDAE
(WHITES & YELLOWS)

RIODINIDAE
(JUDIES & PUNCHES)

NYMPHALIDAE
(BRUSH-FOOTED BUTTERFLIES)

LYCAENIDAE
(BLUES) (SIGH!)

Match the right butterfly family with your balding pattern! (Mine is a Lycaenid. Sigh!)

Sunday Mid-Day, 12th December, 2021

If I were a wagtail...

If I were a wagtail,
I wonder how it'd be

To have a heart so full of cheer,
And such infectious energy

A pair of zestful feet,
And a spirit so wild & free

Would I wag my restless tail
Or would the tail wag me?

Roundglass Sustain, 13th December, 2021

BEHIND THE SCENES OF FLORICAN COURTSHIP

Male floricans must perform elaborate high jumps to woo their mates! Unfortunately, the species has recently been uplisted on the IUCN as critically endangered, because of the destruction of its grassland habitat.

Roundglass Sustain, 18th December, 2021

In December 2021, Buzzfeed News broke the news of a string of sexual offences at the Smithsonian Tropical Research Institute in Panama. Here are some usual 'predators' women come across in the streams of ecology and conservation biology in India. This comic was created with inputs from female ecologists and film-makers.

The Hindu Sunday Magazine, 19th December, 2021

BIDEN AT GLASGOW:

BIDEN BACK HOME:

Sunday Mid-Day, 22nd December, 2021

5 INDIAN BIRDS RECENTLY UPLISTED ON THE IUCN RED LIST:

THE GREEN IMPERIAL PIGEON & THE NICOBAR IMPERIAL PIGEON: NEAR-THREATENED, COURTESY OF HABITAT LOSS

THE MOUNTAIN HAWK-EAGLE: NEAR-THREATENED, THANKS TO DEFORESTATION IN THE HIMALAYAN FOOTHILLS

WHERE IS THE IMPERIAL TREATMENT WE WERE PROMISED?

WHAT'S NEXT? TURNING ME INTO A 'MOUNTED' HAWK-EAGLE FOR THE MUSEUM?

THE FINN'S WEAVER: ENDANGERED FROM DESTRUCTION OF TERAI MARSHES

THE LESSER FLORICAN: CRITICALLY-ENDANGERED BECAUSE OF DISAPPEARING GRASSLANDS

THE 'FINN-ISH' LINE SURE SEEMS CLOSER.

WOW. THERE'S OFFICIALLY MORE GRASS ON MY CREST THAN IN MY HABITAT!

—RMAN

Roundglass Sustain, 20th December, 2021

12 SUSTAINABLE NEW YEAR RESOLUTIONS WE CAN BORROW FROM BACKYARD WILDLIFE:

ECO-FRIENDLY FASHION
-The Bagworm Moth Larva

RECYCLING TAKE-OUT PACKAGING (INSECT EXOSKELETONS)
-The Trashline Orb Spider

BIODEGRADABLE ART MATERIAL
-The Signature Spider

BEING A COMPOSTING CHAMPION
-The Long-flange Millipede

MAKING THE MOST OUT OF FOOD WASTE
-The Black Soldier Fly

DITCHING PLASTICS FOR EARTHENWARE
-The Potter Wasp

ORGANIC HOME DECOR
-The Leaf-cutter Bee

DITCHING THE AIR CONDITIONER TO COOL OFF IN MUD HOLES
-The Brahminy Skink

JUDICIOUS USE OF HOUSING SPACE
-The Carpenter Bee

CASHING IN ON LOCAL & SEASONAL PRODUCE
-The Short-nosed Fruit Bat

DITCHING YOUR DEODORANT FOR YOUR NATURAL MUSK
-The Green Stinkbug

DITCHING COSMETICS FOR HORMONAL MAKE-UP
-The Common Calotes

The Hindu Sunday Magazine, 24th December, 2021

SOME NEW INDIAN SPECIES DISCOVERED IN 2021 CONGRATULATE OUR BIOLOGISTS:

Sunday Mid-Day, 2nd January, 2022

OUT-OF-THE-BOX ADAPTATIONS OF THE YELLOW BOXFISH

BOLD WARNING COLOURATION TO KEEP PREDATORS AT BAY.

RELEASING A NEUROTOXIN WHEN THREATENED, WHICH CAN BE LETHAL FOR PREDATORS.

A BOX-LIKE, RIGID & ARMOURED BODY, WHICH NOT ONLY ENHANCES DEFENCE BUT ALSO REDUCES DRAG WHEN SWIMMING, AND HAS EVEN INSPIRED AUTOMOBILE DESIGN!

WHY THINK OUT OF THE BOX WHEN YOU CAN BE THE BOX?

Roundglass Sustain, 8th January, 2022

A SALUTE TO THE FLYING MARVEL THAT IS THE DRAGONFLY!

THE FASTEST FLYING INSECT, REACHING UP TO **50 KMPH**!

ITS DIRECT FLIGHT MUSCLES CAN PROPEL IT IN ANY DIRECTION, EVEN BACKWARDS!

CAN FLAP EACH WING IN A DIFFERENT PHASE, ENABLING A VARIETY OF FLIGHT MANOEUVRES!

THE PTEROSTIGMA, A PIGMENTED SPOT ON THE WING'S LEADING EDGE, ACTS AS COUNTERBALANCE, INCREASING FLIGHT EFFICIENCY AND REDUCING RISKS OF A CRASH.

THE HIGHEST SUCCESS RATE WHEN HUNTING: 95 TO 100%!

Roundglass Sustain, 9th January, 2022

Earlier this year, researchers from the Wildlife Protection Society of India sighted a Clouded Leopard at 3700 metres in Nagaland, making it the highest sighting of the species in the world!

Roundglass Sustain, 17th January, 2022

SPOT THE MASQUERADE ON MUMBAI'S SHORES!

THE SCORPIONFISH CAMOUFLAGED WITH THE SAND

THE HAIRY CRAB DISGUISED AS RUBBLE

THE GREEN SEA SLUG PASSING OFF AS MOSS

THE SOCIO-ENVIRONMENTAL CATASTROPHE CALLED THE COASTAL ROAD MASQUERADING AS "DEVELOPMENT"

Sunday Mid-Day, 19th January, 2022

India's 2021 State of Forests Report boasted of an 'impressive' increase in forest cover, of course, with terms and conditions.

The Hindu Sunday Magazine, 23rd January 2022

Roundglass Sustain, 29th January, 2022

The Hindu Sunday Mahgazine, 30th January, 2022

THE COMMON HAWK-CUCKOO'S CALLS THROUGHOUT KONKONA SEN-SHARMA'S 'A DEATH IN THE GUNJ'

A CGI SRI LANKA FROGMOUTH IN APPU N. BHATTATHIRI'S 'NIZHAL'

'ALLAH RAKHA' THE LAGGAR FALCON IN MANMOHAN DESAI'S 'COOLIE'

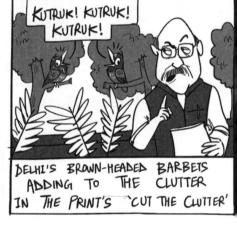

DELHI'S BROWN-HEADED BARBETS ADDING TO THE CLUTTER IN THE PRINT'S 'CUT THE CLUTTER'

What's the unlikeliest virtual birding you have done?

The Hindu Sunday Magazine, 6th February, 2022

ON NATURE TV NEWS TODAY: THE LARGEST NESTING COLONY OF FISH IN THE WORLD, COMPRISED OF THE JONAH'S ICEFISH, HAS BEEN DISCOVERED IN ANTARCTICA'S WEDDELL SEA!

MMM... HIGH TIME SOME OF THOSE ICEFISH WERE IN A WEDDELL SEAL.

Sunday Mid-Day, 6th February, 2022

As the plunder of the Aravallis continues unabated, particularly in Haryana and Delhi, here's a reminder that it is the world's oldest fold mountain range and a treasure trove of biodiversity.

Sunday Mid-Day, 7th February, 2022

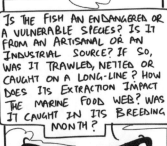

Check out www.inseasonfish.com to know more about the season calendar for Indian fish.
A Valentine's Day special.

Sunday Mid-Day, 13th February, 2022

Gujarat's proposed Rann Sarovar project could well be an impending death knell to several entities: the threatened Asiatic Wild Ass, an endemic species of prawn, the Agariya salt harvesting community and the Little Rann of Kutchh itself.

The Hindu Sunday Magazine, 13th February, 2022

A FEMALE GREATER PAINTED SNIPE COURTS HER MATE

AH, SO YOU'RE THE LADY WHOSE BUMBLE BIO READS "I WEAR MY BREEDING PLUMAGE ON MY SLEEVE!"

THAT'S RIGHT.

SO, LIKE I MENTIONED OVER CHAT, I'M POLYANDROUS. I HAVE NO INTEREST IN RAISING CHICKS, WANT TO MATE WITH AS MANY MALES AS POSSIBLE, AND LEAVE MY BROOD TO THEIR CARE. WOULD YOU LIKE TO BE ONE OF THOSE MALES?

HMMM... WELL... LOOK, I... UMM...

UHH...

PRO TIP, LADIES: WHEN YOU'RE EXPECTING TRICKY CONVERSATIONS, JUST MAKE SURE YOUR LINGERIE HAS FRILLS.

—ROHAN

A Valentine's Day special.

Roundglass Sustain, 14th February, 2022

A World Pangolin Day special.

Roundglass Sustain, 19th February, 2022

SOME INDIAN MAMMALS THAT ARE LIVING FOSSILS:

THE CHOUSINGHA:

THE ONLY ANTELOPE THAT STILL RETAINS A FOUR-HORNED SKULL

THE RED PANDA:

NO CLOSE EXTANT EVOLUTIONARY RELATIVES

THE INDIAN SPOTTED CHEVROTAIN:

AMONG THE MOST ANCIENT LIVING REPRESENTATIVES OF EARLY RUMINANTS

ME:

A DIGITAL MEDIA ARTIST WHO STILL HASN'T PUBLISHED ANY NFTs (SORRY, "DROPPED")

Okay, honest admission: I did try selling a few NFTs months after this comic came out, but my NFT gallery bombed miserably, so I'm back to being a blissful living fossil now.

The Hindu Sunday Magazine, 20th February, 2022

A WHALE SHARK GOES VEGAN

A report by researcher Alex Wyatt concluded that Whale Sharks,
the world's largest fish, and among the few sharks that are filter-feeders,
are actually a lot more vegetarian in their diet than previously thought.

Roundglass Sustain, 22nd February, 2022

The Vaquita's journey has been fraught with unforeseen complexities. US embargos on the Mexican fishing industry have inadvertently been encouraging the spurt of illegal Totoaba fishing (for Chinese markets) further endangering the Vaquita. Mexico's incapability to fund and implement alternatives for fishing communities has now brought the population of the Vaquita down to a dismal ten. This tug of war is only likely to choke the Vaquita further. Has the countdown to the extinction of the world's smallest porpoise begun?

The Hindu Sunday Magazine, 27th February, 2022

SELF·FLAGELLATING DINOFLAGELLATE

Roundglass Sustain, 3rd March, 2022

Environmental Impacts of the Russian Invasion of Ukraine

AVY CONTAMINATION POLLUTION OF UKRAINE'S TER & SEWAGE SYSTEMS

AGRICULTURAL MELTDOWN IN THE 'BREAD BASKET' OF EUROPE, AND THE THREAT TO FOOD SECURITY ACROSS EUROPE & AFRICA

INCREASED RISK OF INDUSTRIAL & RADIO-ACTIVE CONTAMINATION, WITH DONBAS ALREADY ON THE BRINK OF ENVIRONMENTAL COLLAPSE

RISING LEVELS OF RADIATION AROUND CHERNOBYL

RISING RISK OF FOREST FIRES FROM SHELLING & LANDMINES

A MASSIVE SETBACK TO THE CONSERVATION OF 35% OF EUROPE'S BIODIVERSITY, THAT UKRAINE BOASTS OF

DEAR MR. PUTIN, THE WILDLIFE OF UKRAINE WARMLY WELCOMES YOU FOR YOUR FAMOUS ANIMAL-HUGGING PHOTO-OPS ♥

Gocomics, 3rd March, 2022

119

CLIMATE CHANGE WISHES THESE WOMEN A HAPPY WOMEN'S DAY!

AN UNDERGRADUATE IN EPWORTH, ZIMBABWE, DRIVEN TO PROSTITUTION AFTER PROLONGED DRY SPELLS CAUSED AGRICULTURE TO COLLAPSE IN HER VILLAGE...

A YOUNG FISHWORKER IN PANTURA, INDONESIA, PUSHED INTO THE SEX TRADE AFTER FISH CATCHES CRASHED BECAUSE OF UNTIMELY TIDAL FLOODS...

A MOTHER FROM SUNDERBANS, INDIA, COMPELLED TO CHOOSE PROSTITUTION IN KOLKATA, AFTER STORM SURGE FLOODS RAVAGED HER HOME...

KNOW OF ANY CARBON OFFSETS THAT COMPENSATE FOR TRAUMA?

A glance at how climate change is being increasingly linked with rising sex trafficking and prostitution in developing countries that are most prone to climate disasters, on Women's Day.

The Hindu Sunday Magazine, 6th March, 2022

Roundglass Sustain, 9th March, 2022

ELEPHANT SEALS HAVE EXTREME SEXUAL DIMORPHISM, WITH MALES WEIGHING 10 TIMES MORE THAN FEMALES.

Sunday Mid-Day, 15th March, 2022

A MINING GIANT THREATENS ENVIRONMENTAL SECURITY & INDIGENOUS LIVELIHOODS IN AN ECOLOGICALLY FRAGILE LANDSCAPE...

THEIR PUNISHMENT: AN EXTENDED MINING LEASE.

VILLAGERS PROTEST AT A PUBLIC HEARING AGAINST THE LACK OF REPRESENTATION AND A BIASED E.I.A...

THEIR REWARD: A POLICE LATHI CHARGE, ARRESTS, AND CRIMINAL CASES BASED ON DISPUTED CHARGES.

WORD OF THE DAY: "BAUXIT". THE SNEAKY EXIT OF JUSTICE THROUGH A BAUXITE MINING CORPORATION'S BACKDOOR.

On Hindalco's mining operations in Mali Parbat, Odisha.

The Hindu Sunday Magazine, 22nd March, 2022

Roundglass Sustain, 26th March, 2022

Roundglass Sustain, 27th March, 2022

The Hindu Sunday Magazine, 27th March, 2022

Gocomics, 30th March, 2022

The Hindu Sunday Magazine, 3rd April, 2022

Roundglass Sustain, 8th April, 2022

ANTELOPE UNDERWEAR

THE THONG:
WATERBUCK

THE M-STRING:
IMPALA

BLOOMERS:
BONTEBOK

BOXERS:
BLACKBUCK

CROTCHLESS:
TIBETAN GAZELLE

'GRANNY':
HARTEBEEST

SPANX:
ORYX

HONEYMOON:
SPRINGBOK

Gocomics, 12th April, 2022

STREET FIGHTER: MOTHS VERSUS BATS

MOTHS K.O. **BATS**
WITH MY OWL-EYE MARKINGS, I AN OWLET MOTH CAN INTIMIDATE MY PREDATORS WHEN THEY CAST THEIR EYES UPON ME!

MOTHS K.O. **BATS**
I DON'T CARE MUCH ABOUT WHAT I SEE. YOU FORGOT I HUNT USING ECHOLOCATION!

MOTHS K.O. **BATS**
WELL, WELL. MEET THE LUNA MOTH! MY TWISTED TAIL EXTENSIONS DEFLECT YOUR SONAR! AND EVEN IF YOU CATCH HOLD OF THEM, THEY'RE DISPENSABLE APPENDAGES!

MOTHS K.O. **BATS**
AND I, A HAWKMOTH, CAN PRODUCE CLICKS THAT JAM BAT SONAR!

MOTHS K.O. **BATS**
AND I, A NOCTUID MOTH, HAVE EVOLVED SPECIALIZED TYMPANAL ORGANS TO DETECT BAT SONAR!

MOTHS K.O. **BATS**
SAY HELLO TO THE BARBASTELLE! I KNOW HOW TO FOOL MOTHS BY DROPPING MY SONAR VOLUME! SMASH!

MOTHS K.O. **BATS**
AND I, A LEAF-NOSED BAT CAN EMIT FREQUENCIES HIGHER THAN THE RANGE OF YOUR 'SPECIALIZED TYMPANAL ORGANS'! GAME OVE—

MOTHS K.O. **BATS**
NOT SO SOON, TEAM BATS! I, A TIGER MOTH, PRODUCE TOXINS THAT MAKE ME UNPALATABLE. I ALSO WARN YOU OF MY TOXICITY WITH CLICKS THAT DETER YOU IF YOU CHASE ME!

MOTHS K.O. **BATS**
SO, IS THIS A TIE?

UNTIL ONE OF US GETS THE NEXT EVOLUTIONARY CHEAT CODE...

Roundglass Sustain, 16th April, 2022

131

THE WESTERN TRAGOPAN ON LINKEDIN:

STATE BIRD OF HIMACHAL PRADESH
AND OFFICIAL MASCOT OF THE
STATE FOREST DEPARTMENT

THE WESTERN TRAGOPAN ON TINDER:

@crazy_pahadi_loverboy666

Roundglass Sustain, 25th April, 2022

The Hindu Sunday Magazine, 1st May, 2022

Roundglass Sustain, 5th May, 2022

Roundglass Sustain, 6th May, 2022

135

PREGNANCY TIPS FROM THE FRILLED SHARK (THE LONGEST VERTEBRATE GESTATION: 3½ YEARS!)

① COMFORT IS EVERYTHING! GIFT YOURSELF A FRILLY GOWN.

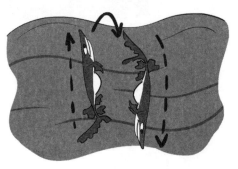

② PROTEINS, PROTEINS, PROTEINS!

③ RECOMMENDED FITNESS ROUTINE: VERTICAL DIEL MIGRATION FROM THE DEEP SEA TO THE SURFACE AND BACK, EVERYDAY

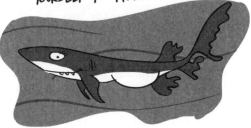

④ YOUR SHARKLINGS SHOULDN'T BE THE ONLY ONES MILKING MOTHERHOOD.

A Mother's Day special.

The Hindu Sunday Magazine, 8th May, 2022

THE MANY ROLES OF A LEOPARD MOM

ASSASSIN

WEIGHTLIFTER

BODYGUARD

STUNTWOMAN

SUPERMODEL

TRAVEL NECK-PILLOW

A Mother's Day special.

Roundglass Sustain, 8th May, 2022

GET YOUR FACTS ABOUT THE BINTURONG RIGHT

ALTHOUGH THEY'RE OFTEN CALLED 'BEAR-CATS', BINTURONGS ARE NEITHER BEARS NOR CATS. THEY'RE RELATED TO CIVETS.

CALLING ME EITHER IS JUST PLAIN **RONG**!

BINTURONGS HAVE SPECIALIZED FEATURES FOR AN ARBOREAL LIFE, SUCH AS A PREHENSILE TAIL, AND WRISTS THAT ROTATE 180°!

DEFORESTATION IS JUST PLAIN **RONG**!

BINTURONGS ARE IMPORTANT DISPERSERS OF STRANGLER FIGS, BEING ABLE TO INCISE THE SEED'S TOUGH COVERING.

APPOINTING ANYONE ELSE FOR THIS JOB IS JUST PLAIN **RONG**!

OTHER THAN DEFORESTATION AND HABITAT LOSS, BINTURONGS ARE ALSO THREATENED BY THE PET TRADE.

THE WORD 'HUMANITY' IS JUST PLAIN **RONG**!

SECRETIONS FROM THE BINTURONG'S SCENT GLAND SMELL LIKE BUTTERED POPCORN.

ASKING ME FOR A CARAMEL FLAVOUR IS JUST PLAIN **RONG**!

A BABY BINTURONG IS CALLED A BINLET.

TWO BINTURONGS MAKE A BINTURIGHT.

A World Binturong Day special.

Roundglass Sustain, 14th May, 2022

Would you also like the Himalayan heatwave cut from Goral and Sons Hairdressers?

Roundglass Sustain, 17th May, 2022

AMAZING ADAPTATIONS OF THE SUN BEAR

AN EXPERT CLIMBER, THE MOST ARBOREAL OF ALL BEARS

EXCEPTIONALLY LARGE CANINES TO CRACK TREE BARKS OPEN AND EXTRACT INSECTS & LARVAE

THE SUN-YELLOW CHEST PATCH, OFTEN USED AS A THREAT DISPLAY

AN EXCEPTIONALLY LONG TONGUE TO EXTRACT LARVAE & HONEY

NOTED FOR ITS INTELLIGENCE, AND THE ABILITY TO MIMIC EXPRESSIONS: AN ADVANCED COMMUNICATION TRAIT!

THIS IS YOUR EXPRESSION AFTER READING THIS COMIC...

AND THIS WOULD BE YOUR EXPRESSION IF YOU TOO WERE THREATENED BY HABITAT LOSS OR POACHED FOR YOUR BILE.

Roundglass Sustain, 24th May, 2022

LIFE LESSONS FROM LEECHES

① PERSISTENCE

② RESOURCEFULNESS

③ THE URGE TO EXPLORE ABSOLUTELY UNCHARTED TERRITORY

④ HOSPITALITY

⑤ A PERMANENT PREFERENCE FOR PUBLIC TRANSPORT, EVEN IN BAD WEATHER

⑥ ETHICAL CODE OF CONDUCT DESPITE BEING A PARASITE

Roundglass Sustain, 25th May, 2022

141

LIFE LESSONS FROM THE BASKING SHARK

① THERE'S NO TASKING WITHOUT A LITTLE BASKING.

② YOU CAN CHOOSE A DIGNIFIED SILENCE EVEN IF YOU HAVE A BIG MOUTH.

③ BE AN EFFICIENT FILTER-FEEDER. FILTER OUT ALL THE NEGATIVITY AND RETAIN JUST THE JUICY PLANKTON.

GULP

④ A SMALL BRAIN SIZE DOESN'T MEAN YOU'RE DIM. IT MEANS THAT YOU'VE EVOLVED FOR A LAID-BACK LIFE.

—ROHAN

The Hindu Sunday Magazine, 2nd June, 2022

World Environment Day outreach campaign comic for ATREE, 5th June, 2022

MOST FISH RELEASE THEIR EGGS & SPERM DIRECTLY INTO THE WATER, LEAVING THE FATE OF THEIR PROGENY TO MERE FLUKE. BUT NOT ME.

I, AN OVIPAROUS CATSHARK, ENSURE THAT MY KIDS START THEIR LIVES IN STYLE, IN CHIC EGG CAPSULES CALLED 'MERMAID'S PURSES'!

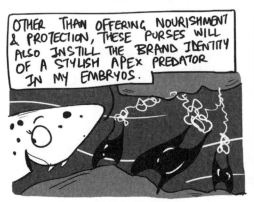

OTHER THAN OFFERING NOURISHMENT & PROTECTION, THESE PURSES WILL ALSO INSTILL THE BRAND IDENTITY OF A STYLISH APEX PREDATOR IN MY EMBRYOS.

OH, AND BEFORE YOU GREENIES JUMP DOWN MY THROAT FOR PROMOTING CONSUMERISM, LET ME INFORM YOU THAT MY PURSES ARE ENTIRELY ORGANIC AND BIODEGRADABLE!

YOU KNOW, I'VE THOUGHT ABOUT STARTING AN INSTAGRAM ACCOUNT, BUT THEN I'D FINISH THE CAREERS OF SUSTAINABLE FASHION INFLUENCERS OVERNIGHT.

The Hindu Suday Magazine, 13th June, 2022

My sincere apologies if buttered popcorn will never be the same for you again.

Roundglass Sustain, 17th June, 2022

Snow Leopards live life on the edge, literally!

Roundglass Sustain, 22nd June, 2022

A Father's Day special

Roundglass Sustain, 1st July, 2022

BEING A BLUE-STRIPED FANGBLENNY

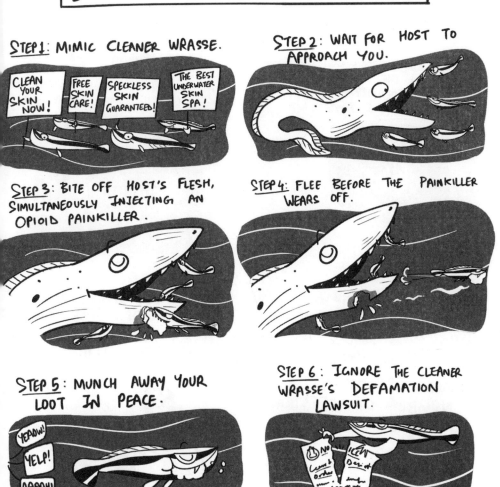

STEP 1: MIMIC CLEANER WRASSE.

STEP 2: WAIT FOR HOST TO APPROACH YOU.

STEP 3: BITE OFF HOST'S FLESH, SIMULTANEOUSLY INJECTING AN OPIOID PAINKILLER.

STEP 4: FLEE BEFORE THE PAINKILLER WEARS OFF.

STEP 5: MUNCH AWAY YOUR LOOT IN PEACE.

STEP 6: IGNORE THE CLEANER WRASSE'S DEFAMATION LAWSUIT.

Roundglass Sustain, 2nd July, 2022

149

4 FASCINATING FACTS ABOUT THE PAINTED LADY BUTTERFLY

THE MOST WIDESPREAD BUTTERFLY SPECIES IN THE WORLD, FOUND ON EVERY CONTINENT EXCEPT SOUTH AMERICA & ANTARCTICA.

MAKES LENGTHY MIGRATIONS IN TEMPERATE REGIONS, TRAVELLING AS FAR AS BETWEEN BRITAIN AND THE MEDITERRANEAN!

FACTORS RESPONSIBLE FOR THE SPECIES' SUCCESS INCLUDE POLYGYNOUS MALES, AND THE HIGH BIOTIC POTENTIAL OF FEMALES, PRODUCING A LARGE NUMBER OF OFFSPRINGS.

COMPLETELY SECURE OF THEIR MASCULINITY, MALE PAINTED LADIES ARE QUITE PROUD OF THAT TITLE.

Roundglass Sustain, 4th July, 2022

TAXONOMISTS NEED THERAPY!

Parastratiosphecomyia Stratiosphecomyioides (A SOLDIER FLY)

WHAT A BRIGHT IDEA TO ASSIGN A FLY THE LONGEST ZOOLOGICAL NAME IN THE WORLD!

Turdus maximus (TIBETAN BLACKBIRD)

DO I LOOK LIKE A... SIGH, FORGET IT.

Lanius isabellinus (ISABELLINE SHRIKE)

HEMPRICH & EHRENBERG NAMED ME AFTER THE COLOUR OF QUEEN ISABELLA'S USED PANTIES, WHO SWORE NOT TO WASH THEM UNTIL SPAIN WAS FREED FROM MOORISH CONQUEST. NEVER TRUST A TAXONOMIST WITH YOUR LAUNDRY.

Dermophis donaldtrumpi (DONALD TRUMP CAECILIAN)

I BEG OF YOU. MAKE TAXONOMY GREAT AGAIN!

-ROHAN

The Hindu Sunday Magazine, 17th July, 2022

VULTURE POPULATIONS IN RAIGAD, MAHARASHTRA, ARE BOUNCING BACK THANKS TO 'VULTURE RESTAURANTS': CATTLE CARCASSES PLACED AT SPECIFIC VULTURE FEEDING GROUNDS BY RESIDENTS.

Roundglass Sustain, 18th July, 2022

On Tanzania's forceful eviction of the Maasai people from their ancestral lands for the creation of a game reserve for the UAE royal family

The Hindu Sunday Magazine, 19th June, 2022

New research (Gitanjali Katlam et al) confirmed the presence of plastics in elephant dung last year.

The Hindu Sunday Magazine, 24th July, 2022

Roundglass Sustain, 25th July, 2022

On the discovery of Meet the Meghalaya thick-thumbed Bat *Glischropus meghalayanus*, discovered and described by Uttam Saikia, Gabor Csorba and Manuel Ruedi.

Roundglass Sustain, 27th June, 2022

THE TIGER'S 'FALSE EYES' ARE HYPOTHESIZED TO BE USED FOR WARNING AND SIGNALLING.

A World Tiger Day special

Roundglass Sustain, 29th July, 2022

The Hindu Sunday Magazine, 31st July, 2022

LEAF ART BY VARIOUS ARTISTS

COMMON MORMON CATERPILLAR

PLAIN TIGER CATERPILLAR

GEOMETRID MOTH CATERPILLAR

WEAVER ANT NEST

COMMON TAILORBIRD NEST

SMALL TREE FROG NEST

KATYDID EGGS

LEAFCUTTER BEE

LEAFCUTTER BEE WITH O.C.D

Roundglass Sustain, 1st August, 2022

THE NEW NORMAL:

ANNUAL HEATWAVES

THE NEWEST NORMAL:

HEATWAVES WITH NAMES

THE NEXT NEW NORMAL:

EXCLUSIVE HEATWAVE EDITION COUTURE

AND MORE NEW NORMALS:

HEATWAVE TRAVEL EXPERIENCES

The Hindu Sunday Magazine, 8th August, 2022

SOME ELEPHANT GESTURES AND WHAT THEY MEAN

CASUAL EAR FLAP

"BEHOLD THE WORLD'S MOST SUSTAINABLE AIR CONDITIONER!"

J-SNIFF

"MMM. FIGS. YEAH, FIGS! OOOH YESS, FIIYYAGGSS!"

TAIL-RAISE + FOOT SWING

"YOUR DEODORANT SMELLS LIKE HUMAN BEINGS. PLEASE GET LOST."

TRUNK-WRAP

"WHO'S A GOOD BABY? YOU! AND WHO'S GOING TO MESS WITH BABY TODAY? THAT'S RIGHT— **NOBODY.**"

STANDING TALL WITH EARS OUTSTRETCHED

"GO AHEAD, PUNK. MAKE MY DAY."

HEAD-SHAKE WITH EARS FLAPPING

"I'M GOING TO CHARGE ON THE COUNT OF THREE, AND YOU'RE GOING TO CALL AN AMBULANCE. READY?"

FOREHEAD NUDGE

"YOUR PLACE OR MINE?"

TRUNKS ENTWINED

"I LOVE IT WHEN YOU SAY EVERYDAY IS WORLD ELEPHANT DAY."

—ROHAN

A World Elephant Day special.

Roundglass Sustain, 12th August, 2022

Freya the Walrus was 'put down' by Norway's fisheries department, as she apparently posed a threat to visitors who would not stop harassing her.

Gocomics, 16th August, 2022

Gocomics, 17th August, 2022

The Hindu Sunday Magazine, 18th August, 2022

COMPLETELY RANDOM INTERJECTIONS FROM NATURE IN MY LIFE THIS WEEK

MY BROTHER ON THE PHONE HEARING THE NEIGHBOURHOOD GREY-BELLIED CUCKOO'S MONSOON YODELLING BEFORE HE HEARS ME.

A PEAFOWL'S SUDDEN CRY CAUSING MY PET SRISHTI'S PUPILS TO DILATE IN AMAZEMENT AS SHE HEARS IT FOR THE FIRST TIME.

MY PET SAKSHI RUNNING OUT TO INSPECT A LOUD SCREECH, EXPECTING A CAT BUT MEETING A MASSIVE INDIAN FLYING FOX!

MY WIFE & I INTERRUPTING OUR MORNING WALK TO INSPECT A DEAD FROG, ONLY TO FIND OUT IT'S A PAINTED KALOULA PUFFED UP IN DEFENCE!

A LEAFCUTTER BEE FLYING WITH A LEAF FROM THE GARDEN STRAIGHT INTO A HOLE IN MY DOOR, MAKING ME GAPE IN WONDER AND FORGET ABOUT MY COFFEE.

A SHORT-NOSED FRUIT BAT'S DROPPINGS ON MY BALCONY CHAIR KEEPING ME FROM SITTING ON IT, NOT BECAUSE I'M TOO LAZY TO CLEAN IT, BUT BECAUSE I WANT THE SEED DISPERSER'S AUTOGRAPH TO REMAIN THERE.

A Nature Gratitude Week special.

Roundglass Sustain, 18th August, 2022

Roundglass Sustain, 19th August, 2022

SKIP A DATE IN STYLE WITH THESE CRAFTY EXCUSES FROM WILD ANIMALS:

THE MUGGER

DINNER? I'D HAVE LOVED TO, BUT I NEED ANOTHER FORTNIGHT TO DIGEST MY LAST MEAL.

THE DESERT JIRD

SEE YOU FOR A DRINK? WELL, I ACTUALLY MEET ALL MY FLUID REQUIREMENTS FROM THE MOISTURE IN GRASSES & TUBERS.

THE SEA SPONGE

THE LAST TIME WE HUNG OUT, I WAS A MOTILE POLYP. NOW, I'M A SESSILE ADULT.

THE INDIAN ROCK PYTHON

SWALLOWED A GODDAMN BOAR LAST NIGHT. CAN BARELY MOVE.

THE LESSER MOUSE-EARED BAT

THE NUMBER YOU'RE CALLING IS IN TORPOR. PLEASE LEAVE A MESSAGE.

—RUAN

Roundglass Sustain, 26th August, 2022

SHIP PROPELLERS... WHRRR

SIESMIC OIL SURVEY AIR GUNS... BOOM!

MARINE CONSTRUCTION... CLANK!

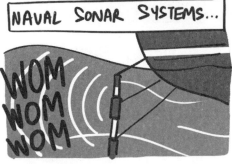

NAVAL SONAR SYSTEMS... WOM WOM WOM

HOW DIFFICULT IS IT TO UNDERSTAND THAT HEAVY METAL IS NOT EVERYONE'S KIND OF MUSIC?!

WOM WHRRR BOOM CLANK

Marine noise is a growing threat worldwide, to the ecology of dolphins and other cetaceans that rely on sonar for communication, navigation and feeding.

The Hindu Sunday Magazine, 28th August, 2022

As devastating climate-change induced floods hit Pakistan, COP vows were rendered meaningless as another developing nation paid in human lives for the mistakes of the developed world.

Gocomics, 31st August, 2022

Roundglass Sustain, 3rd September, 2022

In her Mongabay piece, environment journalist Supriya Vohra outlines how the loss of marine biodiversity and livelihoods has been accelerating because of the Adani–Kerala government Vizhinjam International Seaport. Since then, protests from the fishing community have intensified, and even painted a religious colour by the Kerala government.

The Hindu Sunday Magazine, 4th September, 2022

River beautification is not river conservation. The same goes for lakes and other water bodies.

The Hindu Sunday Magazine, 12th September, 2022

Roundglass Sustain, 23rd September, 2022

TAXONOMY OF INDIAN CONSERVATIONISTS

① Newbieum optimistus
HAS JUST SCORED AN NCBS MASTERS SEAT. ALL SET TO SECURE THE ENVIRONMENTAL FUTURE OF INDIA SINGLE-HANDEDLY. HAS A CAT NAMED J.VIJAYA.

② Fieldworkerum quadripedalis
CRUSHED BY THE WEIGHT OF ACADEMICS AND FIELD EQUIPMENT, HAS NOW EVOLVED TO WALK ON ALL FOURS.

③ Conservationista midlifecrisisum
HAS LEFT HER PROLIFIC FIELD DAYS FAR BEHIND TO TAKE CHARGE OF ORGANIZATIONAL MATTERS. HAVING MORE PENDING GRANT APPLICATIONS TO WRITE THAN GREY HAIR ON HER HEAD, GIVES HER A STRANGE SOLACE.

④ Neglectibus indigenosus
LEADS PROTESTS AGAINST AN INDUSTRIAL GIANT IN HER REMOTE VILLAGE AMID DEATH THREATS. REJECTED AS AN EARTH CHAMPION AWARDS NOMINEE TO AVOID SPONSORSHIP CONFLICTS.

⑤ Neglectibus legalensis
FILING HIS 150th LITIGATION, THIS TIME AGAINST ILLEGAL CONSTRUCTION AT A RAMSAR WETLAND. ISN'T EVEN CALLED A CONSERVATIONIST.

⑥ NGOheadens wornoutinii
EASILY IDENTIFIED BY HER UNUSUALLY LONG ARMS, RESULTING FROM BEING PULLED APART CONSTANTLY IN TUGS OF WAR BETWEEN TRIBAL COMMUNITIES AND THE FOREST DEPARTMENT.

⑦ Communicus switchorus
DISILLUSIONED BY SCIENCE, SWITCHED CAREERS TO BE A SCI-COMM WRITER. NOW, DISILLUSIONED BY WRITING, IS SWITCHING INTO A PODCAST HOST.

⑧ Documentarius deglamouri
A ONE-WOMAN CREW SHOOTING A DOCUMENTARY ON ENDANGERED FROGS OUT OF HER OWN SAVINGS, AFTER THE ROYAL CHARLES GRANT REJECTED HER FILM FOR HAVING NO TIGERS.

⑨ Veteranus GPSii
A RECIPIENT OF NUMEROUS AWARDS FOR HER CONTRIBUTIONS TO CONSERVATION BIOLOGY, SHE NOW WEARS AN HONORARY RADIOCOLLAR AWARDED TO HER BY TIGERS.

⑩ Veteranus questionabilis
PUTS HIS UNMATCHED EXPERIENCE TO THE COUNTRY'S USE BY SIGNING ENVIRONMENTAL CLEARANCES AS A GOVERNMENT RESEARCH INSTITUTE HEAD. FRAMED PRINTS OF MoEFCC TWEETS ADORN HIS OFFICE.

@MoEFCC: INDIA RANKS #5 IN EASE OF DOING BUSINESS!

⑪ Veteranus obscurus
A PIONEER WHO ONCE LED THE FIRST EVER RESEARCH ON A CARNIVORE GENUS. HAS NOW JOINED THEIR PACK BECAUSE OF HIS LACK OF PROFICIENCY IN HUMAN SOCIAL MEDIA.

As a two-part series in The Hindu Sunday Magazine, 23rd September, 2022

Roundglass Sustain, 30th September, 2022

175

New to India, the Cheetah seeks advice from a grassland veteran, the Striped Hyena.

The Hindu Sunday Magazine, 2nd October, 2022

WESTERN LEADERS AT COP 27

The Hindu Sunday Magazine, 9th October, 2022

GUESS WHAT! INDIA'S DUGONGS JUST GOT THEIR FIRST CONSERVATION RESERVE - IN TAMIL NADU'S PALK BAY!

THIS 448 SQ KM PROTECTED AREA WILL HELP CONSERVE SEAGRASS BEDS, CRITICAL FOR OUR SPECIES THREATENED BY HABITAT LOSS...

THE RESERVE WILL ALSO FOSTER THE PARTICIPATION OF COASTAL COMMUNITIES IN MARINE CONSERVATION, AND FINALLY SHINE THE SPOTLIGHT ON A MUCH IGNORED SPECIES...

NEXT STOP: A CAMEO IN PONNYIN SELVAN -2.

Roundglass Sustain, 15th October, 2022

Egypt hosted COP27 in November 2022, amid protests calling for better social justice and press freedom.

The Hindu Sunday Magazine, 18th October, 2022

THE SLOTH IN A HAIR CONDITIONER COMMERCIAL

A World Sloth Day special.

Gocomics, 20th October, 2022

STINGLESS BEE PROPOLIS (A BINDING RESIN MADE BY MIXING THEIR SALIVA WITH PLANT RESIN) HAS MEDICINAL PROPERTIES.

STINGLESS BEES POLLINATE COMMERCIALLY IMPORTANT CROPS LIKE COCONUT, JACKFRUIT, AND A VARIETY OF LEGUMES.

IT'S PROPOLIS™, MIND YOU.

IT'S NOT JUST POLLEN THAT'S RIDING ON THOSE POLLEN SACS. YOUR FOOD SECURITY TOO.

STINGLESS BEE HONEY HAS VERY POTENT MEDICINAL PROPERTIES, RICH IN ANTIOXIDANT FLAVANOIDS.

ALL THIS MAGIC HAPPENS IN WELL-CAMOUFLAGED HIVES BUILT IN UNASSUMING LOCATIONS LIKE TREE HOLES & WALL CRACKS, MUCH LIKE A SECRET LABORATORY!

STINGLESS, BUT NOT ZINGLESS!

M.PHARM, BATCH OF 2022

Roundglass Sustain, 21st October, 2022

Roundglass Sustain, 28th October, 2022

A ban on the fishing and bycatch of Greenland Sharks by the NAFO
raises hope for the slow-breeding species, the longest living vertebrate on earth!

The Hindu Sunday Magazine, 30th October, 2022

COP 27 ended with nations celebrating the long overdue inclusion of a loss and damage fund in the COP agenda. While governments applauded the move, organizations around the world felt that it was too little, too late.

Cartoonathon for the COP27 Resilience Hub and Global Resilience Partnership at Sharm-al-Shaikh, Egypt, 8th November, 2022

BIRTHDAY GIFTS FOR THE 8 BILLIONth BABY

SMOG-THEMED CRAYON SET

FLOOD-RAVAGED DOLL HOUSE

GEOTHERMALLY MINED METAL BATTERY CAR

HUNGER INDEX RANK TOY TEA SET

TOY STETHOSCOPE, OXYGEN MASK & CYLINDER THAT ACTUALLY WORK

The Hindu Sunday Magazine, 20th November, 2022

Nearly 100 species of elasmobranchs (sharks and rays) were awarded protection in the form of an agreement on regulations on finning and trade, between signatories to the CITES (Convention on International Trade in Endangered Species), at the CITES conference held in Panama City in November 2022.

Roundglass Sustain, 29th November, 2022

A Moose from Canada and a Dorcas Gazelle from Egypt discuss the Biodiversity COP and the Climate COP in puns.

The Hindu Sunday Magazine, 4th December, 2022

WHAT'S AT STAKE AT THE BIODIVERSITY C.O.P?

THE DASH OF HONEY ON YOUR PANCAKE

YOUR MORNING COFFEE, POLLINATED BY APIS BEES

YOUR FRIDAY NIGHT TEQUILA PLANS, SPONSORED BY THE AGAVE-POLLINATING LONG-NOSED BAT

THE FIGS IN YOUR ICE-CREAM, FERTILIZED ONLY BY THE FIG WASP

THE ROBIN SONG YOU WOKE UP TO AFTER SNOOZING YOUR ALARM

THE 'FISH' BIT OF 'FISH & CHIPS'

DURIAN, POLLINATED BY FLYING FOXES

CHOCOLATE, POLLINATED BY THE CHOCOLATE MIDGE

THE SMILE ON YOUR PRETTY FACE

Why the Biodiversity COPs should matter to you and me.

DW News, 13th December, 2022

'30x30', a conservation model that was agreed upon internationally at the Biodiversity COP15 at Montreal, was met both with enthusiasm and applause by governments, as well as criticism by indigenous representatives.

The Hindu Sunday Magazine, 18th December, 2022

MOHAMMAD RAFI WAS A MAGPIE ROBIN!

UNMATCHED MELODIES, AS SPECKLESS AS THEIR BLACK & WHITE SUITS

AN INCOMPARABLE REPERTOIRE THAT SERENADES THE NATION DAY AFTER DAY

ABSOLUTELY EFFORTLESS NAVIGATION BETWEEN OCTAVES

THE SQUABBLING COUPLE NEXT DOOR DECLARES CEASEFIRE WHEN THESE TWO SING!

—RAMAN

Roundglass Sustain, 24th December, 2022 (Mohammad Rafi's 98th birth anniversary)

NEW DISCOVERIES OF 2022 CONGRATULATE INDIA'S BIOLOGISTS

THE APATANI SKINK FROM ARUNACHAL'S TALLE VALLEY

TALLE-HO, BIOLOGISTS!

Microhyla nakkavaram, A FROG NAMED AFTER THE INDIGENOUS NAME FOR NICOBAR

MICRO FROG, BUT A MACRO BREAKTHROUGH, BOTH IN HERPETOLOGY & TAXONOMY!

FOUR NEW JUMPING SPIDERS!

THAT'S ONE LONG JUMP IN INDIAN ARACHNOLOGY!

FOUR NEW AZOOXANTHELLATE CORALS FROM THE ANDAMANS

DIDN'T MISS ZOOXANTHELLAE AT ALL, BECAUSE WE TEAMED UP WITH SOME FABULOUS ZOOLOGISTS!

THE GOLDEN SHIELDTAIL SNAKE REDISCOVERED FROM KERALA AFTER 142 YEARS!

THE GOLDEN SHIELD FOR RESEARCH GOES TO INDIA'S HERPETOLOGISTS!

THE PAINTED LEOPARD GECKO FROM ODISHA

BOTH GECKOS AND BIOLOGISTS DESERVE SOME BIG-CAT-STYLE GLAMOUR!

THE LISU WREN BABBLER FROM ARUNACHAL

HONOURED TO FEATURE IN THIS MOMENTOUS CHECKLIST, OR IF I MAY, 'CHECK-LISU'!

Atree, A NEW GENUS OF WASPS, NAMED AFTER THE 'ASHOKA TRUST FOR RESEARCH IN ECOLOGY AND THE ENVIRONMENT.'

CAN'T WAIT TO WELCOME THE INSTITUTE'S 2023 ENTOMOLOGY BATCH AS THEIR NEW PhD GUIDE!

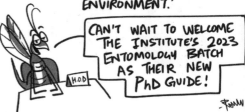

Roundglass Sustain, 30th December, 2022

NEW YEAR RESOLUTIONS FOR BETTER MENTAL HEALTH, FROM WILD ANIMALS

LEARN TO VALUE SLEEP

- Arctic Ground Squirrel hibernates 8 months straight!)

GROWTH THAT MAKES YOU SHINE IS GROWTH SANS ANY DEADLINE

- Jewel Beetle Larva (can take upto 50 years for metamorphosis!)

SHORTER DESK-TIME, LONGER WALKS

1.5 MILLION STEPS!

- Caribou, with migration round-trips of 1200 km!

CULTIVATE A CALMING HOBBY

- Asian Small-clawed Otters that juggle pebbles for amusement

MORE CO-OPERATION, LESS COMPETITION

- Harris's Hawks, that hunt in packs

FOLLOW INFLUENCERS WITH RESPONSIBLE LIFESTYLES

- Jerboa, that has given up drinking and obtains moisture from desert plants

GROW A THICK SKIN FOR DEFENCE AGAINST TROLLS

- Pangolins, that have scaly keratin as armour

PLEASE PICK BODY-POSITIVE ROLE MODELS

Northern Elephant Seal

Mazumbai Warty Frog

Bald Uakari

DW News, 31st December, 2022

ACKNOWLEDGEMENTS

Depending on what sort of thing each of these friends, family members, collaborators and well-wishers is into, I owe them my gratitude, a session of social grooming and allo-preening:

1. Non-humans: Srishti, Sakshi, Chandni, WallE, the sparrows breeding in my hand-made bird boxes, the magpie robins, purple-rumped sunbirds, brahminy starlings and koels visiting my bird baths, and the numerous spiders, insects, reptiles, amphibians I share my home with, for all the inspiration and fodder.

2. Humans: (Late) Sulabha, Ashit and Rohit Chakravarty; Rithika, Roanit and Tina Fernandes; Rushika Banerji, Bittu Sahgal, Anuradha Kulkarni, Priya Thuvassery, Bijal Vachharajani, Sejal Mehta, Kripa, Indu Harikumar, Meghaa Gupta, Rupal Sarin, Vaishna Roy, Rosella Stephen and the Sunday Hindu Magazine team, Megha Moorthy, Neha Dara, Radhika Raj, Julie Belmont, Radhika Suri, Neha Mishra, Swati Thiyagarajan, Craig Foster, Jade Schultz, Annette Hertwig, Samantha Vuignier, Rebeka Ryvola, Pablo Suarez, Arushi Thapar; Sukrita, Bala, Satyabrata and Pronit Lahiri, Jaya Peter, Vinatha Vishwanathan, Michael Scholl, Rekha Raghunathan, Romulus Whitaker, Janaki Lenin, Pritha Dey, Anjali Parate, Dipsha Kriplani, Dhwani Chandel, Jay Kulkarni, Kuhu Majumdar, Tarique and Swati Sani, Ushma and Reema Patel, Neha Sinha, Prerna Bindra, Akanksha Sood Singh, Arati Kumar-Rao, Asha De Vos, Premanka Goswami and Saloni Mital.